Peacock and Quail

A Bird Book for Kids™

By Novare Lawrence

Nada Bindu Publishing Co.

The contents of this book previously appeared in the digital-only editions *Peacock: A Bird Book for Kids* and *Quail: A Bird Book for Kids*.

First Print Edition – January 2016

ISBN-10: 1-63307-011-5
ISBN-13: 978-1-63307-011-0

Published by:
Nada Bindu Publishing Co.
Carson City, NV 89703
Website: www.nadabindupublishing.com
Email: inquiries@nadabindupublishing.com

CONTENTS

Peacock   1

Quail   27

Peacock

Peacocks are considered by many people to be the most beautiful birds in the world. A peacock can lift up his tail feathers to create an elegant fan-shaped display of long, frilly soft feathers dotted with a colorful pattern of "eyes" or spots. These long feathers are called covert tail feathers because they cover and protect the shorter tail feathers underneath.

The long covert tail feathers of a peacock

The proper name for these birds is Peafowl. The Peacock is the male peafowl. He is the one with the long beautiful tail feathers. The female peafowl is called a Peahen. Peacocks display their tail feathers like a giant fold-out fan to attract peahens and as a territorial display to rival peacocks. Peahens can also display their shorter and less colorful tail feathers to challenge a rival peahen or to signal her chicks of a threat.

A Peahen and Peacock walking together

There are three species of Peafowl, each belonging to its own geographic area. First is the Indian Peafowl which is found throughout India. The Green Peafowl is native to Southeast Asia from Burma down to Java. Lastly, the Congo Peafowl is native to the Congo region of Africa. Congo Peacocks do not have long, brightly colored feathers like the Indian and Green Peacocks and much less is known about their life in the wild. On the other hand, Indian and Green Peacocks have been transported and raised around the world for zoos and private estates and are therefore widely known.

The male Indian and Green Peafowl

Indian Peafowl, also called Blue Peafowl, have slender feathered crests on their heads. The male has blue and green feathers covering his head and neck, and bronze-green feathers with black markings along his back. His primary wing feathers are crimson and black. The female's feathers are more subdued with dull green, brown and grey colors. Unlike the male, the female does not have long covert tail feathers. Both male and female have a white patch of skin above and below their eyes.

An Indian Peacock and Peahen

The Green Peafowl of Southeast Asia has a slightly different look. The male has a crest but his neck is mostly green colored. His outer wing feathers are blue and black which cover red or brownish primary feathers. His tail and back feathers are green, blue and golden. The peahen's colors are almost the same as the male although her covert tail feathers are shorter. Both have blue, white and yellow skin patches around their eyes.

A Green Peahen and Peacock

One striking variation in coloring is the white Peafowl. As an adult, its feathers turn completely white although it is not an albino. Albinos are found in all animal species, including humans. Albinos have no coloring in their eyes, skin and hair because they lack pigment in those cells. White peacocks, however, have normal coloring in their eyes and skin.

A White Peafowl

While not very common in the wild, people who own and breed peacocks have found the white peacock to be quite popular because it is so different from the usually colorful peacocks. The white peacock also has a beautiful display.

A White Peacock displaying tail feathers

The Indian Peacock is the national bird of India. They live in deciduous forests where the trees lose their leaves with the changing seasons and also during dry times. Indian Peafowl can easily adapt to farmed areas and villages where water is available to them. Peafowl are omnivorous; they will eat seeds, insects, fruits and berries, small reptiles and even small mammals.

Indian Peacocks gathering together

In Southeast Asia, the Green Peacock is considered to be endangered, which means that their ability to survive in the wild is threatened. The main threats are hunting and the loss of land where they can live. Green Peafowl prefer to live in forested areas but can also be found on grasslands and farmed areas where there is water. They eat fruits and berries, insects, reptiles, frogs, and rodents. They will even eat poisonous snakes.

A Peacock on a thatched roof

The peacock is considered to be a large bird, especially compared to chickens, turkeys, and pheasants to which they are related. A Green Peacock adult male weighs 8.5 to 11.0 lbs. (3.8 to 5 kg) and tends to be much bigger than the adult female. An Indian Peacock male is a bit heavier, weighing 8.5 to 13 lbs. (3.8 to 6 kg) and may be more than twice the weight of the Indian Peahen.

A Peacock foraging for food

Although peacocks are quite colorful and beautiful with their long tails and crests, their relatives, including pheasants and chickens, are surprisingly colorful too. For example, the Golden Pheasant also has long tail feathers and an amazing pattern of white, blue, orange, red and yellow colored feathers.

A colorful relative of the Peafowl, the Golden Pheasant

The male peacock's special covert tail feathers are very long, but they are not the longest among birds. The Crested Argus, which belongs to the pheasant family, has the longest tail feathers but they are not brightly colored like the Peacock's. The Argus' tail feathers can reach up to 5.7 feet (1.7 m) in length. The Green Peacock has tail feathers that are only 5.2 feet (1.6 m) in length at most while the Indian Peacock's are a bit shorter.

Peacocks have the second longest tail feathers

Peacocks are able to fly, even with their long tails, but they don't fly very far or for very long. They spend most of the day on the ground in small groups to be near food and water.

An Indian Peacock flying with his crimson flight feathers extended

Peacocks will fly into trees, or on to buildings or other elevated places to escape danger or to see their surroundings better. They also prefer to roost off the ground at night for safety.

Peacocks fly to buildings or trees for safety and to roost at night

Peafowl are social birds and tend to live in groups for foraging and protection. Males group together, especially during breeding season, in what are called *leks* but may forage on their own when not breeding. When a peahen approaches a lek, the peacocks compete for the peahen by displaying their beautiful tail feathers. The feathers form that famous fan-like shape and the peacocks shake them forward and backward in a graceful dance. They may also offer food to the peahen.

An Indian Peacock displays for a Peahen

Peafowl may live up to 25 years, especially when they are provided food and water, like at a zoo. Peafowl become interested in finding a mate when they are 2 to 3 years old. Peacocks in the wild do not mate for life and a male may have several peahens in his harem. Peahens usually make their nests on the ground but have been found to nest on buildings or in other birds' nests in trees if the nest is large enough.

An Indian Peahen nesting on the ground

Green Peahens lay 3 to 6 eggs while Indian Peahens may lay up to 8 eggs. Only the female sits on the eggs to incubate them and they hatch after about 28 days. Peafowl chicks are called Peachicks up until their first birthday. As with chickens and other related birds, peafowl chicks are able to see and walk the same day they hatch. This allows them to quickly leave their ground nest with their mother. She leads her peachicks around, teaching them how to find insects, seeds and other food, and how to avoid dangers.

A Peahen and her Peachicks

Peahens are very protective of their peachicks. A Peahen will gather her chicks under her feathers to protect them from harm. Peachicks have been known to climb on their mother's back and stay there as the mother flies up into a tree for safety.

Indian Peachicks stay close to mom as they learn to find food

Peachick feathers may vary in color from yellow and brown to a light tan or sandy beige. As they age, new feathers with the adult colors begin to grow. Peachicks from white peafowl parents have yellow feathers. Their white feathers grow in as they get older. As peachicks get bigger, they rely less on their mothers for finding food and for protection. They then live with the social group until they are old enough to look for a mate.

Two Peachicks practice displaying their tail feathers

Peacocks, like other birds, regularly molt. Molting is when old feathers fall off and are soon replaced by new feathers. During the process of molting their primary flying feathers, birds are unable to fly. This can be a dangerous time for them since they can't easily fly away from predators. Peacocks molt their long covert tail feathers once a year, after mating season is over. The new tail feathers are generally longer and fuller after each molt until the peacock reaches full size around the age of 6.

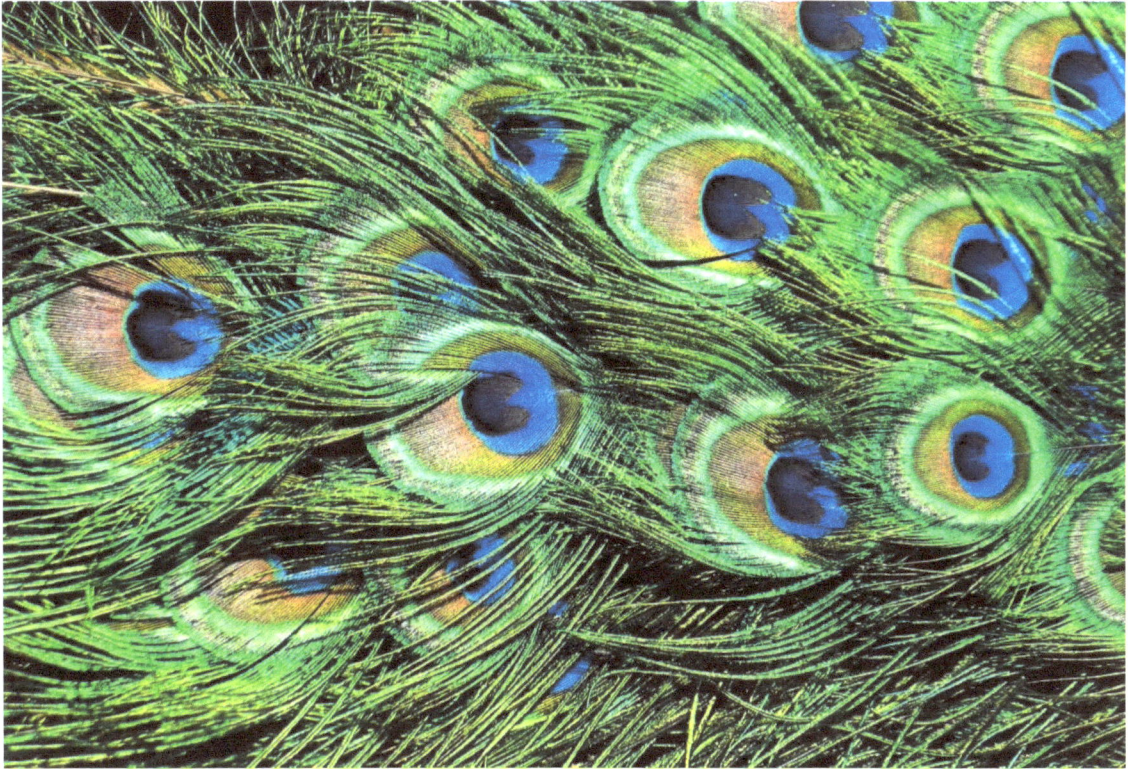

When Peacocks molt, it's the best time to gather their old tail feathers

Peacock feathers are prized for decoration, display and even fashion purposes. Molting is the best time to get peacock tail feathers. A peacock will often drop all its covert tail feathers in a single day. While the peacock doesn't need them, the feathers retain their beautiful colors and shapes for us to enjoy.

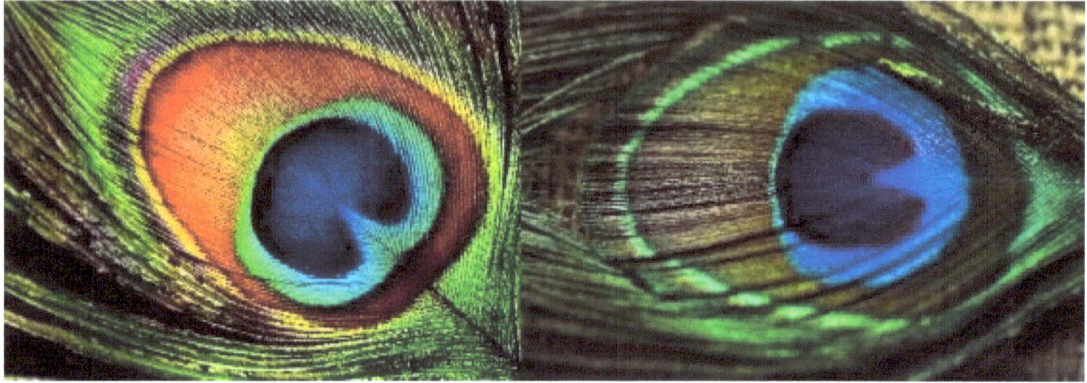

An Indian and a Green Peacock's feathers up close

With such richly detailed and colorful feathers, the peacock has been associated with glamour, money and privilege, especially in the western world due to their exotic Indian and Asian origins. Peacocks have been the subjects of painters and artists around the world for centuries. In their native lands of India and Asia, peacocks are often linked with royalty and deities. Western art also associates peacocks with elegance and beauty.

A Peacock decoration used on a building and a porcelain tea service

Peafowl's cries can be loud although some people like their exotic, echoing calls. They have at least six different calls that are a type of warning or alarm. Both females and males in their social groups will cry out. As one peafowl calls, its cry may be picked up and repeated by one or more nearby peafowls for several minutes. Males also tend to cry out at dawn, like a rooster, and at dusk when they fly to a tree or a rooftop to roost for the night.

An Indian Peacock calling out

Whether or not people like their calls, the one undeniable thing that makes a peacock a joy to behold is its unique combination of patterns, colors, beauty and grace. Peacocks are indeed a beautiful marvel of nature.

The beautiful Peacock

Quail

Quail are distinctive looking small to medium sized birds. There are many types of quail grouped into two major species known as Old World and New World quail. There are 13 types of Old World quail and over 30 types of New World quail, which includes related birds called bobwhites and partridges.

A Bobwhite Quail

Quail have a plump, pear-shaped body with the head at the small end and feet at the big end. They tend to have short tail feathers as most quail do not fly very much. Their feathers are colored to match their typical surroundings with lots of brown, black, grey, and tan colors. These colors along with some white feathers form patterns that make them hard to see in the long grasses, shrubs and forests where they live.

A plump Brown Quail

Old World quail are related to the pheasant family of birds. One of the largest is the 11 inch (28 cm) tall Snow Mountain Quail from New Guinea in Indonesia. Old World quail are found throughout the world while the New World quail are native to the Americas, from Canada down to Brazil. A couple types of Old World quail are able to fly longer distances and usually migrate with the seasons but other quail live in the same region year round.

A Harlequin Quail, one of the Old World quail species, is found in Africa

New World quail are most familiar to people living in the U.S., Canada and Mexico. They do not migrate and spend most of their lives on the ground. They will often hop and flap their way up to a higher place like a fence or tree branch to get a better view of what's around them. They only fly short distances as a last resort to get away from predators.

A Gambel's Quail, one of the New World quail species, on a short flight

Another example of the Old World species is the Japanese Quail which is one of the few quail that migrates. From Japan, Russia, Korea and China where they live most of the year, these quail fly to Southeast Asia and southern China in the winter. The males are between 3 and 3.5 ounces (90 – 100 g) in weight while females are slightly bigger than the males.

A Japanese Quail

One example of the New World species is the Scaled Quail, named because its breast and back feathers look like scales. Sometimes it is called a blue quail due to the bluish-grey tint of its feathers. Its crest resembles a small white tuft on its head so this quail may also be called a Cottontop. Scaled Quail live in the United States, from Arizona up to Colorado and across to Kansas and Oklahoma. Adults weigh from 6 to 6.5 ounces (170–184 g) and it's the largest New World quail at almost 12 inches (30 cm) tall. They have been known to live in groups with as many as 150 other quail.

A Scaled Quail checking its surroundings

There is another species of bird called the Buttonquail. While there are sixteen different types of buttonquail, they are not related to the Old World or New World quail. They look like quail when you see the shape of their bodies and they also live mostly on the ground but their beaks are shaped differently and are longer. They are genetically different from true quail and live in Asia, Africa, Australia and parts of Europe.

A Barred Buttonquail looking very much like a true quail, except for the beak

When we think of birds, we usually think that birds fly. That's why they have wings. Quail, however, are only one of many types of birds that spend most, if not all, of their lives on the ground. There are lots of other birds that share this trait. From large ostriches that can't fly at all, to peacocks, chickens and roadrunners, these birds spend most of their time on the ground.

Like quail, the roadrunner, chicken and ostrich also spend their lives on the ground

Since most quail spend their time on the ground, that's where they sleep, build their nests and look for food. Most of their day is taken up by foraging. Quail eat insects, seeds, some plants and even roots, depending on the seasons and the climate where they live. Their territories are usually within a few miles of a reliable source of water too.

A King Quail, also called a Chinese Painted Quail, resting on the forest floor

Birds that fly look for their food in different ways from terrestrial or ground-living birds. Many of them catch their food while flying. Small flying birds can snatch an insect right out of the air. Large birds like hawks and eagles can catch smaller prey from the air, ground or water using their talons without stopping. Sea birds will fly up first and then dive down into the water to catch fish below the surface and then take off into the air again with their fish.

A Pelican with a fish and an Eagle about to catch one

Quail are social birds and like to live in large family or community groups called a covey. While foraging for food, one or two quail will often find a high place to watch over the rest of the group. The sentries will sound an alarm if they spot any danger so that the group can run for cover. If they have to fly away, they wait until the last possible moment, then, they all take off at the same time, going in all different directions. The noise and distraction this causes helps to startle a predator which allows the quail to get away safely.

A Bobwhite Quail up on a fence acting as a sentry

Quail don't run as fast as a roadrunner or an ostrich, but they are very good at disappearing into bushes, shrubs, grassland and other ground cover to escape from predators. Once inside cover, they stay very quiet and, helped by their feather colors and patterns, they easily stay hidden until they know the danger has passed.

A Bobwhite Quail blends into the ground cover

Quail build their nests on the ground. With so many varieties of quail, there are lots of differences in the number of eggs and in incubation times for different quail. While many quail lay from three to six eggs, bobwhite quail may lay from ten up to fifteen eggs. Quail chicks hatch between 16 and 30 days later.

A nest of Quail eggs

Like many other ground-living birds, quail chicks hatch with their eyes open and downy feathers grown. Chicks can run within an hour or so after hatching so that their parents can lead them to safety if any predator comes near their nest. The chicks learn what to eat and how to stay safe by following their parents and the covey. As they get older and grow bigger, adult colored feathers grow in to replace their downy feathers.

Japanese Quail chicks

Birds that fly tend to nest in high places like trees and cliff ledges. The parents need to find a safe nesting place because their chicks can't walk or fly after they hatch. The parents must bring them food until they grow bigger and stronger and can fledge, or leave their nest and fly on their own. Even then, the parents often continue to feed their chicks even after they fledge while they teach them how to find food for themselves.

Baby sparrows grow up in the nest - - Grey Owl owlets grow up on a rocky ledge until they can fly

Quail mature quickly and for species like the King Quail, they are able to produce eggs after they are one month old. Some quail lay eggs in several spots and then roll them along the ground to bring them together into one nest. When living in a covey of quail, some females may lay their eggs in another female's nest and let that female incubate them. Usually quail chicks are raised and watched after by the entire covey, not just by their parents.

A well hidden ground nest

Some species of quail are considered to be game birds. This means that they are hunted by people as food. Quail are also farm-raised just like chickens but chickens are larger, can lay bigger eggs and feed more people. This makes raising chickens much more practical for farmers but adult quail and quail eggs are both considered to be a delicacy by many people.

Young quail being raised on a farm - - Spotted quail eggs next to three larger chicken eggs

Quail live around five to six years in the wild although this can vary some among the different quail species. They are vulnerable to various predators, particularly when they are young. Hawks, owls, foxes, coyotes, and even snakes may hunt quail or seek out their nests for eggs. To avoid predators, quail will hide in thick brush or grassland, listen for alarm calls from other quail and, if necessary, take a short flight to get away.

Can you spot the bird hiding in the grass?

One of the favorite activities of quail is a daily dust bath. Living on the ground as they do, quail are exposed to lots of insects. While they like to eat insects, some insects are small enough to get under their feathers. To stay healthy and clean, quail can take several dust baths each day. They often do this as a group since most quail live together in a community.

A Quail couple take a dust bath

Small birds like sparrows and much bigger birds like ostriches also love a good dust bath. The dust is very fine and helps to remove insects, scrub the skin and keep their feathers clean. And it's just a lot of fun.

Size doesn't matter when a sparrow or ostrich needs a good dust bath

For all the species of quail, the most fun to see in the wild are those with the special tufts of feathers on their heads. This is called a plume or crest. It is formed by a group of feathers that droop over the quail's head. For species like the California and Gambel's quail, both the male and female birds have a crest.

A male and female Gambel's Quail

Other bird species also have crests. The peacock's crest is distinctive with its many separate thin feathers extending from the back of its head in a line with little fluffs on the ends. The common crane features a different type of crown where the circle of feathers resembles a pin cushion surrounding their head. The Cockatoo has yet another type of crown -- a group of wider feathers that normally lay flat but can be raised above its head if it is excited.

A peacock, a common crane and a cockatoo also have crests

There are many species of quail around the world that come in different sizes, colors and patterns. But the California quail is a favorite as most people think that it's very cute. This quail makes a soft "pit-pit" noise as one of several calls it uses to communicate with other quail while foraging. It also has an alarm call that sounds like "chi-ca-go." California quail have a variety of calls and songs that they sing. The male will often sing along when his mate sings.

An alert male quail

Whether it is standing guard by itself or running along with its covey, the distinctive crest of feathers makes the California quail very easy to identify. If you're ever in the western United States, in the hills or grasslands, be sure to look, and listen, for this attractive and distinctive bird.

A cute California quail

ABOUT THE AUTHOR

Novare Lawrence loves researching and writing books about Nature. She shares the knowledge and beauty of our natural world with kids young and old hoping that we will all do our part to help preserve our planet and all the wonderful species upon it.

You may learn more about her books and the *A Bird Book for Kids™* series at her website:

ABirdBookforKids.com

And at NadaBinduPublishing.com

A Bird Book for Kids™ Books   by Novare Lawrence

Digital:

Print:

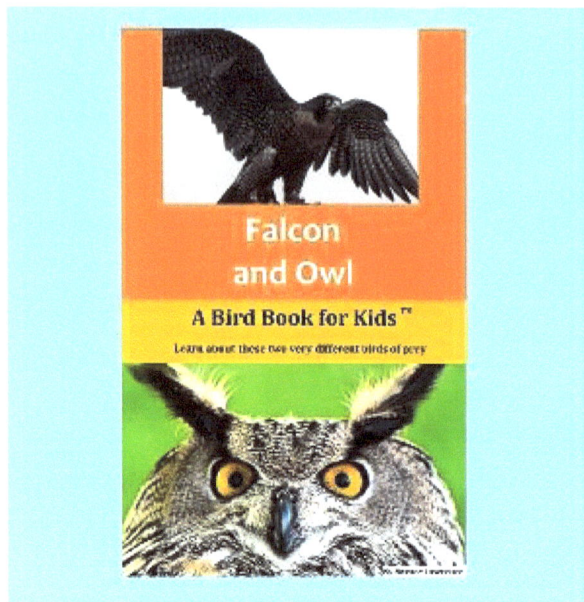

Falcon
and Owl

A Bird Book for Kids™

Learn about these two very different birds of prey

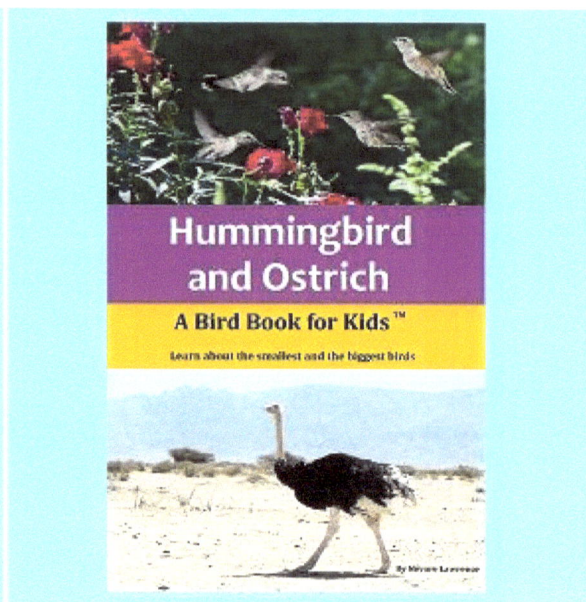

Hummingbird
and Ostrich

A Bird Book for Kids™

Learn about the smallest and the biggest birds

By Steven Lavrenar

www.ingramcontent.com/pod-product-compliance
Lightning Source LLC
Chambersburg PA
CBHW060817270326
41930CB00002B/66